Grab the Horn

My Life and Saga

Grab the Horn

My Life and Saga

As Recollected by Mike Hotz

Michael Hotz

Grab the Horn

By Michael Hotz

Cover Images:
Front cover: Mike Hotz
Back cover: Military Medals Awarded to Captain Michael V. Hotz

- Bronze Star
- Purple Heart
- Defense Meritorious Service
- Air Medal
- Army Commendation
- Army Good Conduct
- National Defense
- Vietnam Service
- Vietnam Gallantry Cross with Palm
- Vietnam Wound
- Vietnam Campaign
- Combat Service
- Air Combat
- Vietnam Gallantry Cross – Unit Citation
- Vietnam Defense
- Tet Campaign
- Cold War Victory
- Airborne and Air Assault
- Foreign Expeditionary Service
- Overseas Service
- U.S. Army Service
- Armed Forces of U.S. Honorable Service
- Armed Forces Retired
- Armed Forces Expert Marksman

Interior photos: from the collection of Michael Hotz

734 East Aspen Avenue
Fruita, CO 81521

Ordering information: Quantity sales. Special discounts are available on quantity purchases by book clubs, corporations, associations, and others. For details, contact the publisher at the address above.

ISBN 978-0-932624-44-4

1. Main category—[Biography] 2. Other categories—[Memoirs]—[Survival]

Dedication

My life hasn't been simple but it's sure been interesting. I want my kids and grandchildren to know some of what I went through. So, this is dedicated to my sons Marcus and Hunter, my daughters Ciara and Cheri Mariah, and my grandchildren, Luke, Ben, and Lily. My message to those future generations is simple: reach out and grab the horn—in other words, whatever you decide to do, do it well and hang on!

I don't know if God goes to church,
but I know He goes to war.

— Dr. Barry L. McAlpine

A photo from my childhood days, growing up on the DeBeque ranch; shows my parents along with my younger brother, Perry, and me displaying a couple of kittens

CHAPTER 1

Early Memories

September 1948 to May 1966

Mark Twain once said a person should never ever tell a lie, but he added that it's okay to embellish a story. The names and dates here are the best of my knowledge, but I am sure I will have to edit time and time again.

I was born in September 1948, in Grand Junction, Colorado to Vernon Hotz and Judy Hotz. I grew up on a farm in DeBeque, Colorado, in Bonita Valley. (Some people put a space in the name DeBeque, but I don't. It was the last name of my coach and music teacher who was the grandson of the founder of DeBeque. So, the town is named after the family that founded it, and I use the way the family spelled it.)

My dad came from Pagosa Springs, Colorado, from an area that was called Devil Creek Ranch. My father was born in 1923. He married Judy, whose first name was Pauline (she hated that name her parents gave her.) I have one brother, Perry Hotz, who was born in 1950.

As the oldest son, I learned to ranch very early in life. After chores, I grew up riding milk cows as a way to practice for rodeos. Going through pictures of my folks, my first memories as a child were of my dad working on a bean farm and we boys, my brother and I, came home a little rattlesnake. I was reprimanded for that. And I once had to go get old Ed Willams' false teeth that had been dropped into the well. I was lowered into that small dark well with no light and I felt like I was hundreds of feet down but probably it was only 30 or 40 feet. Anyway, I did retrieve the teeth.

(Ed was claimed to be a notorious horse and cattle thief. But Ed was also our friend and neighbor.)

We returned to Grand Junction the following year. Dad searched for work and ran into a farmer (Bud Kennedy) who made a deal on the farm in DeBeque. (At one time, DeBeque was the largest cattle shipping yard of Colorado!) Dad started farming the land. The deal was Dad was renting to own. While cleaning up the farm, my brother and I burned down the haystack playing with matches. After quite a paddling, my brother and I sat on the roof watching the neighbors come and help put out the fire. That was a great way to move into a new town.

Eventually, Dad bought the farm from Bud Kennedy and then he bought the neighbor's farm from the widow and a couple of years later, Dad bought the farm across the road.

I recall that the school bus would come to the highway in front of our house and take us to the school which was three or four miles away. Lots of stories about that highway. My first grade had 11 students. The entire school had about 112 kids. Since I was shy, I fell into a group of local ranchers' kids. I would stay at night with different kids and likewise.

Once we kids got to be 8 or 9 years old, we got horses. My friend Pete had a Welsh, blacky or dark.

Another friend, Harold, had a couple of horses. I had only a workhorse, that we used for mossing the ditch, until Dad bought me a small buckskin at a stock sale. I named him Tony. It turned out that Tony was a gone-sour roping horse. At a full run, he would go to bucking. Since I had no saddle, I would come back leading him, humble, and he became quite famous among my friends. Everyone wanted to top him off but without a saddle, nobody ever did stay on him. We had another good horse named Dagwood, but I used him for handling cows and rodeos.

Another memory pops up. We had a little white pet goat. I named him Goat, and we rode him. Goat would climb on people's cars and skid off. He would butt and chase our friends. He was our pet, just like a dog. Unfortunately, he also used to break into our house and eat Mom's plants. Eventually, my mother grabbed a .22 rifle and shot the goat. My mother tried to serve us goat meat for supper, but nobody, including my Dad, would eat it.

Sometimes I would stay with my friend Pete's family. He had a Shetland-cross who would throw everyone off.

When our family had lots of guests over, we didn't have a lot of food. We didn't mind the company. As me and the other kids grew older, we went from horses to trail bikes. The neighbor gave me a moped scooter which you had to pedal to start. Mine ended me up in a ditch and I didn't like bikes after that.

In May, all the folks around had an annual rodeo. We kids started barrel-riding, and we rode bucking milk cows. The older kids rode bulls. I do remember that a father of one of the younger girls got drunk and fought one of the tourists. I sat on my horse and watched that.

Every Fall, Dad would take me out of school for two weeks to help him with farm chores. I was 9 to 11 years old somewhere in there and, by that time, I was used to doing farm chores. At home, I would get the sheep in, also the horses and cows, lambing sheep, feeding and milking the cows, and setting out salt licks.

Later, Dad bought me a Heiser rough-out saddle. At the time, it was an expensive saddle, so I got it in place of my summer wages. Over time, as I got older I advanced to raking and cutting hay and driving an ensilage truck (a '39 Ford—two or three-ton) several miles to a ensilage pit. I also started doing farm work in our fields with a 9N Ford tractor and, within a year, I got promoted up to a 600 Jubilee Ford for mowing. Dad continued to run the baler, and I was following Dad running the corn cutter plus doing livestock chores and irrigating and stacking hay bales or driving the tractor that pulled the slip. I worked during summer vacation both days and nights.

We worked hard on the farm. Dad would come in at 6:00 in the evening for supper and then go out again until after dark to finish the chores. He worked for the neighbors and shortly after bought an adjoining farm. After that, my workload increased. I worked after school. Dad worked nights at the feed mill.

In the fifth grade I broke my right arm, so I was out of school for two weeks. Everyone from school came down to visit me in the hospital. The school bus brought two grades down 30 miles for an hour with cards and get-well wishes. It was great! I was there probably two weeks and finally went back to school. Anyway, I couldn't do my regular chores, so Dad had to pick up the slack.

I was spending time with my friends at night, sometimes on stayovers. We had work to do ,but we got a chance to spend some time together. With Harold, Pete, and Ned at first. At the end with Lynn.

One year they had a dance at the Community Hall for the adults. Everyone would get together and swap wives for dancing and to drink and fight. My parents didn't drink. My mom made her own wine—chokecherry and dandelion wine. But no one would drink it except us kids.

As a kid, I used to climb out of the window while my brother was feeding chickens and scare him. When he told my parents, I would deny it.

A few times my dad's brother would visit from Moab, driving an old Chevy car, and the next tiime a brand new one as he was a prospector and had many claims in the area for uranium. He operated in and around Charlie Steen. One time Mom's sister brought her kids up—two girls. We would play house. For my brother and I, that was our first real exposure to the female species.

Looking back now and remembering, I'm not that proud at what I did when I stayed at my friend's. The wild deer would come down and eat their crops. There were hundreds of them, maybe five or six hundred. So, the neighbors essentially paid us to shoot or run over as many deer as we could to help preserve what we could of the crops. They even provided us with shells. We did that even to the point where, when we had trucks with bumper guards, we would hit as many as we could.

My dad maintained the irrigation canals. Also Blue Stone Ditch and Larken Ditch by the bridge in DeBeque and we had to go into the river and remove stones to moss and clear the ditch. Harold's foster sister died from drowning in the ditch when she was eight years old. There was a big community effort to find her.

In the school, we had a Spanish teacher who everyone liked a lot. Some of the boys would go to his house, and we found out that he would try to get my brother and I to do things. We said "no" and told my mom, and she went to the school board to report him. When the school took action, he was long gone. Some of the other boys who had gone to his house got questioned and it turned out he had been abusing them.

Finally, I was 16 and driving. I saved all my money stacking and bailing hay. So, I bought an Impala, white with red interior. Ned bought a Mercury Comet. Harold bought a brand new Fairlane 500. Pete ended up with his dad's friend's car which he hated because it was an old '54 Ford. Pete hated it because it was not as new as our cars. So, he painted it purple, and called it the purple people eater after several wrecks. Didn't bother him that he wrecked it because it was old.

Me having a driver's license caused my dad and I problems because I would come home late and not get my chores done. And I would drink beer under the bridge. I remember one incident when I came home late, chores not done. Dad was in the field when I showed up. He began to yell at me. I had a few beers, and I decided to stand up to him. I was young and drunk and got scared when he jumped over the fence and came at me. My mom was home, so she stopped any fight before it happened. After that, I went and stayed with friends for a while.

When I still had one year left of school, Dad sold our three farms to an oil company for $65,000. This was a lot of money back then. I hadn't graduated yet, but Dad had decided to go to Alaska and buy a new ranch. So, my folks insisted I go with him.

In the meantime, my dad had aquired the brand X HANGING U which he got originally from a little man named Peanuts who lived in town. The X HANGING U brand goes back to the late 1800s or early 1900s. It's believed to be the old Cleary Armstrong band. Armstrong was quite a notorious bronc rider who could stand off the saddle while the horse was bucking and kick a horse in the belly. The brand probably came out of Deer Park with either Dave Knight or Charlie Chitman. The state board brand inspection information, which later my dad gave to me when I was anching in Gunnison, still hasn't documented the exact origin. I've decided to keep the brand in the family, but I may sell it since probably none of my kids will go into ranching.

When I was 17, I went to renew my driver's license, and we got hit which totaled the front end of my Impala. Several months later, Dad found a '64 SS Chevelle. Dad helped me get this car. That partnership went well. Then one day, being late to school, I took a corner at 90 miles an hour and my brother was with me. I looked up and seen my dad standing there watching me. I heard about that.

In high school, I had a teacher from Oklahoma. Myself and others got interested in his teaching. He taught electronics, which was very popular. Because of what I learned in class, the Air Force tried to recruit me even before I graduated when I was still 17.

I was always interested in sports. In the 6th and 7th grades there was a kind of Peewee basketball and track. They wanted football but the Junior high didn't have enough students for football.

Even with all my chores on the farm, I played basketball as my main sport. As a freshman, I started playing. Our team was in the top spot for the tournament after getting beat at finals the previous year.

Anyway, by the time I got into freshman year there was full-time varsity basketball, and you had to tryout before getting into the varsity. I would play at night at home and with other friends. We played until it got dark rain, snow, or shine. And that extra playing made a difference.

Some of the boys I stayed with had older brothers, including Bobby and Lyle and Coe. So, we would pick out different teams. We got good because, when you're from a small school, you get picked and you get more playing time. I played as a freshman and then more as one or two of the other kids graduated their senior year. By the time I got to be a sophomore my class was pretty much the first five. I think that allowed us more experience and when we were a couple of years older, we got into the playoffs with Riverside High School, which had students from Silt and New Castle.

We had that game won, and we were in the finals. But we partied the night before too much. So, the coach was really disappointed. We lost by just a few points, which was quite an upset. The first five had pretty much fouled out. Anyway, that was my highest point game. I think maybe over 30 points or maybe even 33 or 34.

Some of what I did in high school led up to problems. Too much freedom, being late for chores with the Impala, drinking under the bridge on weekends, and fighting over girls or anything.

After games, we would chase the other team's cheerleaders, who were a lot of times girlfriends, so we ended up in quite a few fights. Different schools would have different dances and likewise, more fights. I remember one time at Rifle, a group of the Rifle boys came up and my friend Ned got into a fight with one. He had Ned on the ground kicking him in the head. I said that's enough and he popped me right in the mouth. After a blank out, my friend told me that I'd thrown his head into the bumper several times so we left. Later, cars from Rifle started showing up at school and I would have to get rides with teachers as he turned out to be a Navy man and me beating him was too much for his ego

When I graduated, we headed for Alaska to buy Dad's dream ranch. Since I didn't want to leave my friends, I stayed in the back of the camper and didn't socialize with my family. But I could do 1000 hand presses. In the camp ground in Alaska dad met an older couple, and the man had talked about having a heart attack and didn't know how to get his new truck back since they had two trucks. And Dad said, take that boy back to his friends, he can drive anything. At the border put him on a Greyhound bus, he's been a pain in the ass.

After working on the ski area for several weeks, playing pool in the afternoons and drinking beer, Bobby's older brother joined the Marines, so we thought it'd be cool to join the Army under the buddy plan. After our minds were made up, we'd drink beer at the station and we shaved our heads, and if any of our friends came by to drink beer, we'd throw them down and shave their heads, which was irritating to the mothers because they still had to take graduation pictures. At Denver, at the induction, we met our civilian platoon who were of every color, hippies at the time. So, a Sergeant came up to me and jerked my hat off and said, what are you, a smart ass? Take this group to Fort Bliss, Texas. And he handed me a large manila envelope. I said, why me sir? I'm not a sir, I work for a living. You're not only ugly, you're dumb. On the trip I lost three or four recruits at a train stop. So we got to Texas and I got chewed out by another Sergeant and sent out as a squad leader. That was the end of the buddy plan as I hardly saw them. Mom and Dad had to come back to sign as I was only 17. Mom was against it, Dad was for it. Thought maybe the Army would shape me up.

Michael V. Hotz, son of Mr. and Mrs. Vernon L. Hotz of De Beque, was commissioned an Army second lieutenant upon graduation from the Engineer Officer Candidate School at Ft. Belvoir, Va., June 23.

CHAPTER 2

Basic Training and More

June 1966 to July 1967

I graduated from DeBeque high school in May 1966 and a month later I entered the U.S. Army. Basic training at Fort Bliss in Texas was eight weeks long. If we made good progress and caused no trouble, we got leave on weekends. And, fortunately or unfortunately, the fort basically is right across the border from Mexico. Several of us, four or five at least, crossed the bridge to Mexico. Walking down into the old part of the town, it was very dirty. Nothing going on except playing pool, drinking beer, and girls working the bars. But I stayed around for after-hours. The girl I met there was probably 18 and Mayala was a pretty little girl. So, we had our thing and the next weekend I planned to come down and do the same thing. The rooms were 8 feet x 8 feet with a gunny sack for a door. But within about a week we got transferred so I never did see her again. I had no way to communicate. I guess you're going to say: lost love of the night. Yeah.

I reported to my next post at Fort Leonard Wood, Missouri, in October 1966. It was damn cold there. My MOS (military occupational specialty) was combat engineer, so I had to work explosives, build bridges, and also tear bridges down. We used rafts, so I was always getting wet and cold. Training there lasted another eight weeks. By that time, I didn't want to go to Vietnam because I realized what it was. And not knowing anything about what to do, I asked my Sergeant if I could stay out. And he said no way. You're a combat engineer and you're going to go right over to Vietnam within about two weeks, unless you want to go to OCS. I asked what's OCS? And he said it's Officer Candidate School. It's one long and hard six months for an enlisted man to become an officer. So, I decided, well, six months of staying stateside is a lot better. So, in the spring of 1967, I reported to Officer Candidate School.

OSC was near Washington D.C. at Fort Belvoir, Virginia. My son Hunter would later be at this same place on January 6, 2021, during the riots at the White House. OCS was pretty tough. It was a lot of marches and running wearing backpacks. You had to run five miles to qualify. I had to do the run over twice because I didn't break the five-mile mark the first time. But the other parts of it I did real well. Along with another friend we both scored 94%, so we became the upper-classmen. As upperclassmen, we would treat the other newbies as bad as we had been treated. So, OCS was a kind of revolving s*** hole.

Six months later, I got out of OCS as a second lieutenant. This made me the youngest commissioned officer outside of the U.S. Military Academy at West Point. There again, I was about to get shipped out. So, again I asked is there any way I can stay stateside? Well, yes, I was told there's two ways. If you come back over, and you qualify, you can fly. Well, that sounded good but unfortunately, I picked helicopters. It was explained to me that observation helicopter flights were high up in the air and they were. They had to fly high to support missions by identifying positions. My primary helicopter training was at Fort Wolters in Texas.

I got to ride with another lieutenant who was headed back east when he dropped me off in Washington D.C.. While I was waiting for the assignment to flight school, I was assigned to temporary duty at

Fort Belvoir as a district attorney for the court system. We were trying AWOLs (military personnel who were accused of being "absent without leave".) Generally, this duty was kind of screwed up because you'd make a deal with the prosecutor on the weekend and then the poor E7 (a non-commissioned officer, in other words, a sergeant) who was getting court-martialed usually bit the bullet and pleaded guilty. But, in one case I remember, the guy was a sergeant, and they were going to bust him to a private because he left AWOL to go see his wife who was having a baby. And so, I argued with the prosecutor. I said it's not going to happen, and in that case, I got that guy off. The prosecutor was pissed at me, so for a while after that I didn't get too many deals.

In the meantime, I think I was there probably a couple of months, and it was kind of an interesting job. While back there, I bought a '67 442 Oldsmobile so I had something to drive. I remember that the girls in the office were interesting. They were flirts. So, I ended up with a thing for a blonde named Patricia.

When I got my next assignment, I got notified that I could be accepted back in Fort Wolters, Texas for primary flight training which was four months. And that was on a Hiller aircraft with a little piston engine. After you solo, you just go out and spend time. I was in West Texas looking around. I looked for deer and just sat around sitting looking at the country.

But the way I did it, I got my hours in and after that I qualified on the solo and then I went into another section which was a Bell Lycoming turbocharged helicopter. For lodging there, I got a notice of living with a flight instructor, not my own flight instructor, but another one. He really wanted to stay by himself, but we had few drinks, and it worked out. I think he must have been a divorcee or something but that lasted a while. I think the total was half the time I was there. So, four months.

I noticed there was not much social activity around Fort Wolters. I think the nearest town was Mineral Wells, Texas. Outside of that, for military guys there was not much going on. After a while, I got orders to go to Fort Rucker, Alabama, which is in Dothan. That was when I actually got to fly Huey's. It was hotter than hell in Dothan. I must have been there in summer time. I got a house with three other guys, Ron and Skip and Wayne. So, we lived there with four guys and an episode every weekend. But somehow, I got stuck with paying for all the utilities. I don't know why they put me in charge, being the youngest. I don't think I volunteered for that one.

When weekends came up, we went to Panama City and rented another house down there for weekend parties. So, it got to be a Friday afternoon road race. People would talk about who could make it to

Panama City the quickest. My 442 and I got in fastest and I won for a couple weeks. I had Sears radials with thin white walls which I painted red. I had pavement marks from outrunning the state patrol. About the race, I think Skip with a Porsche took over and Ron didn't have much of a car. He did have a big bike which I ended up buying. It was Triumph 500 Daytona. Ron got a BSA and then Skip got a BMW. And Wayne, I don't think he did much with a car or a bike but somehow, he got down there and those were big parties in Panama City. The parties lasted all weekend. Different people were showing up all weekend.

I remember one big party there when we drank a bunch of Red Mountain wine that Skip brought from California. It was expensive stuff, two-and-a-half dollars a gallon. And I got educated on wine right then. I drank way too much and woke up on the beach with my face stuck to the sand. A family came along and one of the kids asked, "Mommy is he dead?" I finally moved. I didn't look good. Had sand the size of a plate stuck to my face. After that, I never liked wine, although I later was able to stand the stuff.

Meanwhile, the flight school was tough. There was a lot of mapping weather, learning different gauges, learning the helicopter, and all the pre-flight checks. It wasn't drills and running like the OCS part.

It was a lot of more interesting technical stuff. Eventually, I got to the end of all that and then it was time to get ready for going to Vietnam.

After flight school ended, I was going to take a leave and go home. The class was over, and I had put in a lot of hard work on the choppers, map reading, learning all the instruments, brief lights, and weather. Eventually, I got to where I could land and not spill a cup of coffee. At those times, while we were going along, the instructor would cut the power off and I had to find a place to land without spilling his coffee. If you spilled the cup, you had to go back on Saturday mornings. I got to where I could really be right on the dime.

A couple weeks before going home for a leave, after I was weaned off of the wine for a while, I remember that Wayne and I ended up going out and ran into some locals. And we got into it with them. Wayne wasn't much of a scraper. I was holding my own, but he was getting thrashed. After they left, I took him home. When Ron saw what had happened to Wayne, he said, "Let's go teach them some rules, Hotz." But I told him that was probably not a good idea, although I knew how mean he was when we wrestled on weekends. Ron was a wrestler from California Poly. Anyway, we went back against my better judgement.

Ron was pretty drunk when we went to an after-hours café. I hadn't changed my shirt so I was a bloody mess. When we walked up to the locals' table, they laughed. There were four or five couples sitting there. Ron jumped right in and asked, "Who's the meanest s*** here?" They said, "If you want more, follow us to the rodeo grounds."

On the way out I told Ron, "This is not a good idea." He laughed.

When we got there, Ron jumped out running toward the other guys' car. He came on something like Brutus or Bucko, in other words a wild and tough guy. By the time I stopped our car and got out, Ron had some ringleader on the ground and had his thumb in the guy's eye about ready to pull it out. The guy was screaming but not moving. Ron asked who was next, but they wanted out. So, I didn't get touched. Ron was tough for sure. Another time I was with him and coming out of a bar and he jerked a pretty tall and big palm tree out of the sand and pushed it on some other guy's car. Just being mean.

Finally, I got a couple of weeks to go back to Colorado for a leave. Earlier, I had met some girl while I was still at Fort Wolters. She came out for a visit from Snyder, Texas. She was a schoolteacher and a prim and proper type girl. But anyway, when she visited us, she was appalled at the guys at the house but got over it.

So, she ended up going home to Colorado with me on the leave. I put a motorcycle in a U-Haul trailer and pulled it with my 442. We headed out for Colorado, and I blew the motor up in Dalhart, Texas with that trailer and a bike. So, I got a room for us and then I called my dad. He got ahold of a neighbor who came down with a pickup and picked up the trailer, the bike, and us. I left the car there to get the motor rebuilt and I was going to go back and get it later. It was a long drive home and hot in this neighbor's old GMC pick-up. There wasn't even air conditioning.

We headed back, all four of us in the front seat. The trip was a bit bad for the girl. Deidre was having her period and she kind of had a smell about her, but she was too timid to say anything. I saw my dad and the neighbor Kenny looking at each other.

It was a terrible long trip. Anyway, we got back, and I unloaded the motorcycle and turned the trailer in. In a week or two they were supposed to have the car done before I went back. Meanwhile, Deidre was actually getting pretty serious about me. But I was going to Vietnam. So, I said there was no sense getting involved then.

Eventually, when I got wounded, she came to the hospital. She got really serious but I considered myself a recent cripple and so I shot that down. I wasn't interested in a serious relationship at the time.

My Letters

August 1966 to June 1968

I wrote home, of course. What young kid away from home didn't? I'm looking over that old stack of letters now with only ten-cents worth of postage stamps and envelopes addressed to just:

Vernon L. Hotz
DeBeque, Colorado

No real address and no zip code. Later, I added the zip code but not every time. Somehow the letters made it home. There aren't a lot of letters, but they remind me of what was going on both in the Army and at home. The first bunch came from me as *Private* Michael V. Hotz, a raw U.S. Army volunteer stationed at Fort Bliss, Texas, for basic training. That first letter is dated August 31, 1966. It starts out: "Thought I'd better write and tell you how I'm doing. We arrived in Texas Tuesday evening and didn't leave the reception station until this Monday. We are at our basic training camp now but don't start our 8 weeks training period until next Monday."

I was with two guys I knew, and we all agreed that, except for getting up at 5 a.m. and getting run around like cattle to eat and get issued equipment, we agreed everything has been a snap for us. I asked my folks to write and tell me about everything that was going on, especially any excitement that was going on "without us terrors there." I also asked to see about my brother, Perry, fixing up and selling my red Chevy. In most other letters written while I was still in the states, I continued to ask how the car sale was going. I closed by spelling out the exact (complicated) address they'd need to use to write back to me.

My next letter home thanked the folks and Perry for sending money but said I was okay between poker and interest on loans I'd made to other soldiers. Already in business for myself, I guess. In a next letter to my brother Perry, I mention that I've decided not to go the route of going to an Army school to become a warrant officer. Officer Candidate School (OCS) seems a better idea although the training will be tough. The Army needs junior officers and, by going to OCS, I'll still get a chance to fly and probably in helicopters. In my next letter to Perry, I ask for all the news about high school, who's going with who and what they are doing for excitement. I also ask him to send some things: underwear, shaving supplies, soap, and the pen and pencil set I got for graduation. But not too much stuff because I don't have room to keep it. I'm firm in my decision to go to OCS because it will probably keep me out of Vietnam for a year or so. Although I figure I'll go eventually.

By October 10, I'm definitely going to OCS but it looks like I'll be heading to one of the tougher training programs. I've gained about 15 pounds; up to 180. I just managed to qualify at the shooting range. Had trouble at first because they run you around and make you shoot from unsupported positions. I did better shooting prone but wish I could have hit more targets. I definitely want the folks to sell my cars. Looks like I won't be needing them for quite a while. It doesn't look like I'll get to come home between assignments. My last October letter from Fort Bliss tells about going on bivouac (which I describe as basically a camping trip.) We camped in the desert, had to go through a gas mask drill. I got my mask on upside down at first and "nearly croaked." We went through the infiltration course twice, once in daylight and once at night, crawling about 150 yards with explosions going on all around and machine guns shooting just over our heads. I went before an OCS board where rumors say about 90% of candidates are rejected. I seemed to do okay and feel I am still in the running. Don't have orders yet for my new post.

My early November letter comes from Fort Leonard Wood, Missouri. My friend Gary is here too, and I see him fairly often. I'm still bound for OCS but will be trained here as a combat engineer. I describe that training in detail. Everyone here seems to be older than me. The training is seven weeks including Saturdays and we should be done by Christmas, and I should be getting a leave to come home. Very likely this cycle of training will go to Vietnam, but I still hope to get flight training before I have to go. Expecting snow any day now. A week later, in my next letter I mention a clipping the folks sent me about a wreck involving my friends in which a girl was killed. I suggest some of my friends who are acting up should join the Army. Training is tough, running a mile in the morning, double-timing everywhere, but it will hopefully get me physically ready for OCS which is rumored to be even tougher. We get most evenings off and passes on weekends.

I haven't found time to write, and my December 12 letter tells of a wet bivouac in tents in the rain and wind. Gary and I will be traveling together to come home for Christmas.

In January 1967, I'm back at Fort Leonard Wood. I was sorry I wasted my last night at home. Didn't mention how I spent it, but apparently it wasn't productive, and I ended up not being able to say goodbye to friends. The passes driving back to Missouri were slick and it snowed all through Kansas. Gary and I stayed at Mary Alice and Gale's and got an early start after breakfast. Back at the Fort, it looks like I'll be here until February 12.

My next letter tells about all the harassment and hazing that goes on to weed out OCS candidates who can't take it. The details of all we had to endure still makes me mad, but I close my letter by guessing that I'll get through it and eventually be doing the harassing of others myself.

I thought I'd be at Leonard Wood until February but before January was over I was sent to Fort Belvoir, Virginia, and OCS at the U.S. Army Engineering School. Now instead of Private on my return address, I get to call myself O/C for "officer candidate." I'm getting little sleep, doing everything at top speed (including eating), running two or three miles every day, doing two hours of physical training, standing daily inspections, and generally being run ragged. I'm also busy sprucing up my gear and there's plenty of harassment going on. However, I wrote, "It all sounds worse than it is because you are so busy that you don't have time to think about it." One of my main worries is I'm probably the youngest one here. Candidates aren't supposed to apply for OCS until they are 18½ years old. I applied earlier than that and "I hope they don't get to looking into it because it might cause some trouble." Meanwhile the regular classes are pretty academic, and I did get a chance to briefly tour Washington, D.C. I encourage people to write to me, but I can't promise a prompt response.

I spent three days trying to finish my letter of February 7. In it I brag that I'm getting plenty of physical training and that I did close to 200 pushups in about 15 or 20 minutes. I also like the classes. It's about like being in college. "The training is great so far even if I got out of the service tomorrow I would be able to use a lot of the ideas, and the knowledge." I'll be applying for flight school soon. It will be a tough nine months with classes half of the day and flying the other half. Most of those who don't make it do so because of academics. "If I don't make it, it will still be a great experience for me. Everyone should go through it. It is like the whole Army should be. So far I still have the intention and the confidence to make it through, and time will tell." In other news, I have a chest cold and I'm hoping it doesn't get worse. Send my vitamins. "By the way, I have been going to church. The first time was to get out of details, but now I go to relax and I kind of enjoy it." I close with a story of trying to sneak hamburgers into the barracks only to be discovered and have them confiscated. However, we're going to try and sneak in another 100 hamburgers tonight. "Life is dangerous around here. Ha. Ha."

A letter written late in March apologizes for not writing sooner. I've just returned from Fort A.P. Hill where we did field training. Nice to get away from OCS routine and pressure. I had to qualify with a rifle again and this time I made expert only missing four shots out of 60. I also shot a machine gun, a rocket launcher, and a grenade launcher. "I believe I learned more last week than I have in the past month." I was company executive officer (next in command to the company commander) in charge of meals and transportation. A lot of responsibilities in control of 80 men. I passed my flight test and physicals and I'll be getting a class date for flight school when I graduate here. I made a lot of requests to send treats, especially a German chocolate cake, if possible.

I guess the folks came through because in a letter without an envelope but dated with the year 1967, I thank them for the cake and say it was the best I've ever eaten. The package contained cookies and brownies, and I ate some and passed the rest around. I also shared the cake with friends and had to put the frosting on using a ruler. Sending most of my pay home in a check. I told my folks and Perry to take $10 each and buy themselves something they want for themselves. Don't worry about taking it. Put the rest in my savings. I have a local checking account but keep only $20-$30 in it. Our group must have got food poisoning because 85 or 90 out of 120 ended up going to the hospital last night, me included. There was a big investigation with a lot of brass (officers) checking things out. Might have been some undercooked chicken. We are on 48 hours bed rest and a tactical march will probably be called off. I was supposed to be the executive officer for the exercise. "The Tac officers picked the ones that they thought could do the job, so I felt pretty good about it. I have about an 88% average now too, so thus far I am doing o.k." I'm still the youngest and many others have been in college but I'm still about 3rd or 4th as grades go so far. In addition to academic subjects, we've had "rigging, field fortification, and other things like that." I was worried about map reading and I studied extra by flashlight. I got 85 on the first test, then 93, the second highest in our class. "All that this means is I have things going pretty much my way if I can just keep it up." Three more weeks and we turn upperclassmen. Our upperclassmen just turned red. "That means they only have 3 more weeks until they graduate, so it sure gives you the push to go on when you can see them come that much closer every day." When I see privates and other enlisted men and their appearance I can see the difference in myself. "Now we fall out with clean starched uniforms, high gloss boots, and it makes you feel like you are pretty sharp even if you aren't."

My next letter admitted to a couple of incidents that could have gotten me kicked out of OCS. While we were at A.P. Hill, one of my best friends took a smoke grenade from an exercise and a group of us set it off and filled up an office. We were called before a board and could have been dismissed from OCS. No one was kicked out and I got off fairly lightly with a dozen walking tours (walking back and forth in an area for hours) and a three-week restriction to the post. Turns out our officers and even the tactical officers whose office we smoke-bombed were pulling for us. Later, during a nighttime escape and evasion exercise, my friend and I hid in a truck when the others were let off to try making it through a group of soldiers playing the part of aggressors without getting captured. We jumped off at a stop sign and made it through a swamp but still beat the others back. We didn't actually cheat by riding in the truck since everything was fair to do in the exercise. I got my records evaluated so far and, out of our whole platoon, I was 8th in leadership and 4th in academics. Not bad, but I think I can do better. This weekend we become upperclassmen. We'll have to see how that goes.

In April I sent an airmail letter addressed to my Dad. Not sure what motivated it, but I guess I knew his birthday was coming up in May. It was written at 11 o'clock at night and it's special enough to quote word-for-word:

> *"Dad, Just a short note to wish you a happy year. I hope you go ahead and tackle whatever you would like to do.*

Don't hold back. Fine words for a jughead like me trying to tell you some good advice. I look back, I must have been one of the most stubborn and arrogant of the town. The worst part was that I knew it and could have done something while a lot of the rest just didn't know any better. Well, I really didn't intend to write this type of letter, but I got to thinking and had to translate a little of it on paper. I am sending some more pictures, about the only thing I have to spend money on. Most of the courses are quite interesting and you can get something out of them. You would be surprised how much background I have in some subjects compared to the ones that have actually had college courses on the subject and don't know the first thing about it. This is a messy note, but it is about 11:00 now. Write more later. Bye, Mike (over) Take $25 out of the (enclosed) check and get something that you want or would enjoy, even if you don't need it! I'll enjoy it more seeing you spend it than if I actually went out and blew it."

I continued this theme in my next letter which included a birthday card and my handwritten note "Dad, To a much better friend than I sometimes realized. Your son, Mike." This letter also mentioned that I've achieved upperclassman status (it's called "turning white") and have just started dealing with my "smacks" (underclassmen.) Now that I'm an upperclassman, I'm amazed at how "sloppy, dumb, slow, etc." an enlisted man is before he comes to OCS. I'm the youngest upperclassman here and it's something which those who are 26-29 years old and older address me as "sir" and shiver in their tracks waiting for me to say something. Some are sergeants who have been in the service for years, but they still show me respect, and I show respect for them. My studies are tougher but I'm still holding my average. The underclassmen do my boots and bed, but I still have to spend time studying. I'm their squad leader and I'm with the smacks "from the time they wake up until they are all quietly tucked into their little beds."

My next letter home described an off post pass which I used for a trip sightseeing in D.C. I also mention that I'll be out of OCS by the end of June. It's time to start buying officer uniforms since I'll no longer be an enlisted man. That's going to run into some money, about $200 for some of the greens and blues. Getting my orders to flight school will mean a little extra cash. "I hope so, from the sound of things here about Vietnam it's going to take extra cash to get me to do it. Not really. I think things are a sight safer in the air than the ground. I read in the paper about 800 and some helicopters and 625 planes getting shot down. I don't know when this has been from but probably all total. Most of the pilots are rescued or at least you don't hear about them. O well, enough about this. I have a long time or maybe forever until I go over to the Jolly Green Jungle.

My next letter, written in May 1967, comments on the Virginia weather which is humid compared to home. We went through a "confidence" course yesterday and a few of the obstacles were quite challenging. I took some pictures which I'll send when they get developed. Last night, we went to a party thrown by our Tac (tactical) officers at one of their homes. Six couples, the officers and their wives, eight or ten of us candidates (six having dates) and five or six secretaries and stewardesses. To take to the party, we bought the host a sloth for a pet. Cost $60 and about the "ugliest and stupidest animal I believe I have ever seen." Our Tac said we "are 3rd lieutenants now as far as he is concerned so we should start acting and living like it." Quite a compliment since we're actually about to be 2nd lieutenants. "Next week, we turn red (or Senior) the last phase of the program."

We're starting to buy uniforms earlier than I expected and they will be costing more. $85 each for two greens and $85 for blues which I'll need to get since I'll be going to flight school and will be stateside. So, I'm asking the folks to send my last check back because I'll need the $100. Try to send it soon. Our extra clothing pay won't come until the last of June and I'll have most of my new uniforms bought by then. We should be getting our orders soon and from what's happening to others it looks like those signed up for flight school are "held over here for anywhere from a couple of weeks to a couple of months" as a utility officer. I'm thinking about buying a car maybe. Having a hard time deciding if it's worth it although I'll be nine months in flight school.

My next letter, postmarked July 22, 1967, has a return address for Lieutenant Michael V. Hotz at Camp Drum, near Watertown, New York just below the Canadian border. I joke that the place is like Camp Granada (like in the Allen Sherman song.) It's a "summer camp" for national guard, early release soldiers, and reserves. I'm a Range Maintenance Officer for a network of nearly 50 different shooting ranges from rifle to tank. There are three of us lieutenants, six sergeants, and 32 civilians working for us. Of the ranges, I only have five or six, but they are the farthest away from camp, 31 miles. I have my own jeep and go out to check the ranges almost every day.

I have a car now, a 4-4-2 Olds. Perry will explain what it is. I'm having the title and insurance papers sent home. I quote all the car's features from a magazine article. Coe, who was visiting here and should just about be back home now, took some pictures of the car. There's supposed to be good fishing up here and there's a swimming pool at a lake about two miles away. The Worlds Fair Expo is in Montreal (150 miles away) and maybe I'll get to see it. "Watertown is about 50,000 but an army town so there aren't too many stray girls unless we go to Syracuse or Montreal which is a ways." One of the lieutenants is 23 and the other is 19. The 19-year-old and I get along well with each other and the sergeants, and we are both OCS so we "naturally are a little more on the ball." The other lieutenant is ROTC, and we may have to take over—"at least we have been talking about it. Ha! SOUNDS GOOD. HUH." Eating at the officers' mess and staying at the bachelor officers' quarters (BOQ.) I may have written more from this temporary assignment, but they aren't in the stack of old letters.

Next up in the stack is a January 1968 letter from Mineral Wells, Texas. The envelope is from The First National Bank of Mineral Wells. Probably because I sent the folks a check to pay off the car. In addition to sending the check I apologize for not spending more time with the folks when I was home on leave. I ended up running around, taking people to the airport, feeding a bunch of dogs for a friend, and visiting girls in Grand Junction. I think it might have been better for them to come down where I was. We probably would have seen more of each other that way. My brother, Perry, must have been playing basketball in those days because I advised him to hog the ball and shoot it every time it came close. "Then the country would remember the (Hotz) name because after next year I doubt that they will see much of anybody connected with it." I hope the money I'm sending will pay off the car. I want Dad to keep checking on some land deals and I want some more life insurance before I go to Vietnam.

My next January 1968 letter bears a street address: 1001 SE 19th Street in Mineral Springs, Texas. I imagine I was in flight school at Fort Wolters but that's about four hours from Mineral Springs, so I'm not sure. I'll be flying this weekend to catch up because bad weather put us behind. I heard on the news last night that a hardware store in Grand Junction had to sweep all the snow from in front of their store because they'd sold all their snow shovels. It hasn't snowed much in Texas. More of an ice storm but it still shut things down. I'm asking the folks to send me some civilian clothes: Levis, sweat socks, and so on. We try to go to the gym to work out and play basketball.

I enclosed some pictures showing my drinking mug and little bar. (No idea where those pictures are now.) I've got to end the letter because it's time to go flying. Got to get in 20 hours this next week. That will be 3 or 4 hours a day. Yesterday I only got an hour and 45 minutes flying. My 'copter had a bad clutch, and the transmission fell apart when I started it up. "Almost got a chance to practice a forced landing."

In a loss letter with no envelope and no inside date, I mention the house in Texas and sent some photos. "We started flying cross-country and night flying this last week, but we got hit by another storm so it's us down a little. We go to Rucker March 15."

A March 23, 1968, letter comes on Fontainbleau Motor Hotel stationary although it's postmarked from Dothan, Alabama, which is near Fort Rucker where I'm flying Huey helicopters. The Dothan address is 200 Colonial Avenue. The hotel in is New Orleans where I was taking a brief break from military life. I'm living off-base again with other soldiers and most of this letter isn't about New Orleans. It's about buying furnishing for our off-base house. Our Alabama house is 30 minutes away and are paid 10 cents a mile to drive and $11.80 extra a day for housing. "It's good to get away from the military life at the end of the day although this is a nice post. The club is the best I've seen. It cost $1½ million, completed last June."

At Fort Rucker, I could apply to fly the new Cobra, the new jet helicopter by Huey. It costs 4 or 5 hundred thousand and goes about 250 mph. So far there are only 10 or 20 in Vietnam. “It would cost me another year to fly it, but boy would I like to fly that baby! Of course, I doubt if I get a chance, considering only the top of the class and higher ranking officers get a crack at it. If I got the chance, I would probably grab it like a fool and sign my life away. Be about 23 when I get out; pretty old Huh?” I tell Perry to consider coming down here when I leave to help drive back and “get a feel for the beach. Be a nice vacation for him and even someone else if he wanted. He had better live it up a little before he starts college, get out of the rut for a while.” The trip to New Orleans cost a lot and I’m thinking about buying a bike, so not sure “how much money I’ll send home this month.”

A March 30 letter from Alabama has me talking about finances. I’m sending checks home to put in savings and checking. “Should give me about $1,600, may be able to get another couple of hundred this month. Hope you find some land so I can double my money fast.” I must have mentioned in another letter or a phone call about a girl I was seeing because I mention Mom’s “marriage counsel. Tell her not to worry, the kid (me) is not going to take the big step yet. If a guy was looking for a wife, this would have been a good one." I go on to describe her as a college graduate, French major, with a new car and a lot of class. Father's an oil contractor in England, so must have money. "Too bad I am a confirmed bachelor, Huh?" I asked about them sending clothes in my New Orleans letter and I got them in good shape. As for flying, we are flying instruments only now. "They cover everything so that all you can see are instruments, about 20. You get vertigo, dizzy, sick and disgustingly mad. But I guess we have to do something to earn the money. Won't be long until we start on the Huey, the pride of the army."

A May postcard shows a DC-3 like the one I flew to look at my land. "Think I'll buy a couple of more lots. Liked what I saw. Maybe Dad and Per ought to think about it." Guess this might have been the start of my interest in real estate.

In June 1968, I sent Mom a get well card. It was pretty sentimental with a cross and roses, and I added a note: "Get well fast. Shouldn't worry too much, I guess. Nothing could keep you down for very long. Keep that fighting spirit (or stubborn) that we seem to possess. Love, Mike."

On June 21, I sent another letter asking how Mom is doing. "We are really starting to fly now. Formation, low level, route recon missions. (Even went out and fired one with machine guns and rockets the other day. Fired about 4,000 rounds and was fairly accurate. Need some more practice. I think I'll put in for gunships overseas. Might just as well shoot back if I get shot at." I tell my Dad why I didn't send him a birthday or Father's Day card because I bought something for their house and will have to bring it instead of shipping it. (Can't recall what it was exactly.) Talking finances again. Want to put some money aside to live off of while I'm in Vietnam. I made 1st Lieutenant this week and will be in two years in August. I'll get a base pay jump and a jump in flight pay and sustenance along with hostile fire pay. "Back from Vietnam I'll be captain over 3 years and around $800, $850 a month." Time to fly at 5:00. I'll try to call Sunday sometime in the evening.

1st Lieutenant Michael V. Hotz, 116th Assault Helicopter Company; somewhere near Cu Chi, Vietnam

Chapter 3

Vietnam and Back Again

August 1968 to February 1969

I went back to Texas and got my car by myself, and it was tight but ran good. On the way back to Colorado, when I started up over Grand Mesa, I could feel the car open up. So, when I came down the other side and came to the cutoff, I thought about the old days when us kids would race on that stretch from the town of Mesa to DeBeque. Known as the DeBeque Cutoff, it was considered a shortcut. It was nine miles with lots of turns. Back then, the record time for driving it was 12 or 13 minutes. I had nothing to lose. I knew I was going over to Vietnam, and I thought, if I had a wreck I wouldn't have to go. So, I put the hammer down and opened it up and got that car broke in. I think I made it in 9½ minutes. I don't think that time was ever beat.

Anyway, I didn't wreck, and it was off to Vietnam in August 1968. I had to go to Denver and fly out on Braniff Airlines. We stopped in, I believe it was Hawaii. Anyway, we didn't get off or anything. I remember that when we flew into Tan Son Nhut, which was a Saigon airport, the stewardess came back and said we were getting some fire. So, going in we had to hunch down in our seats. She told us this is pretty common. So, we landed. I don't think we got hit because they just shot rounds overhead, trying to keep the planes away. I met Wayne there and we got on the same plane and went back to a base camp called Cu Chi. It was a pretty big base camp.

My unit was called Assault 116th Helicopter Company and things started right off. I had a couple days getting checked in. After checking in, we started flying right away and I think it was the first or second day that we were dropping troops in a rice paddy. That's when I saw one of my classmates take a bullet through the bubble of his helicopter. It hit him right in the head which just blew apart and splattered the bubble. That was an eye opener.

Then a couple of days later, we were dropping some more troops in about eight choppers, dropping in on a quick drop. I told my pilot, and I think also the crew chief and the gunner to pull up their line of sight. I think I said something like, "The rounds are hitting right in front of us, you dumb s*** that's incoming, duck!"

Well, we had armor plated seats, and we had the vest and the helmet but after seeing that classmate shot through the bubble, nothing was guaranteed.

Anyway, we were flying Cu Chi down and we were an assault helicopter outfit, so we were supporting multiple units: 25th, 101, 1st Cav, Big Red One—anybody who needed support gunships. We were kind of a whore outfit. Our company at Cu Chi had several different MOSs. There was an engineering detachment and some medics and infantry outfits. As I said, it was a pretty big base.

One day, after we'd flown several missions, we went to Vung Tau, and I saw some cement on the deck in bags. Which gave me the idea to trade beer to get a neighboring platoon or company with bulldozer to dig a big hole like an ensilage pit to make a swimming pool. Then we could have the gooks or the South Vietnamese, in other words some friendlies, come in and use cement to block it off on the edges and ends.

Vung Tau was an hour and a half away from our base. It was another base camp but smaller than ours and Navy. I called our Commander and told him to send me all the slicks (transport helicopters) they could gather up. That would take some time, so I started for the forklift to find an operator. I figured each helicopter could start with about one pallet then take off right over the ocean. Then the next one in line behind could come in and get loaded. This seemed like a good plan, and we were loading and flying off until a couple Navy officers came around and asked what I was doing. They told me that cement is Navy property. I said we had an army requisition for it. They asked to see our orders and I told them they're on another ship out there somewhere. So, I turned around and told the pilot to get ready to go and I told the guy with a forklift to load that last bundle.

"I've got to leave," I told him.

He loaded that bundle and also loaded everything that was left. So, we were heavy when we started over the ocean and couldn't get *translational*, which meant we couldn't climb or maneuver very well. Finally, we got up over the ocean and the other airships were in front of us and they had already landed. We did a running landing, so we won't split our skids, and we unloaded everything there next to where we wanted the pool. We had just tarped it when it started raining.

I never did get to see it finished but somebody sent my pictures.

After watching what was going on in Vietnam for three or four months, I decided it was better to be flying up higher. So, I put in for gunships but had three or four warrant officers ahead of me and they didn't want to lose their positions. I had to take a flight test at the time and after six weeks or so my name came up. They picked a day that was pretty hot, and I'd been at the officers' club until closing and I had drunk too much. About 9:00 in the morning, we started circling around and around and around until I ended up throwing up. The puke went back behind and hit one of the crew chiefs and maybe even the gunners. So, yeah, I passed but achieved the call sign *Ralph*.

Okay, back to missions. We started flying all three areas: Tay Ninh, Cu Chi, and Vung Tau/Tonsa Nute. That was a lot better because we were a lot higher and we always went in ahead of slicks. So, that was my next two months, and it was my routine until I got hit with a mortar attack.

It was nighttime on November 6, 1968, when the base was attacked. I remember getting thrown on the floor. I was bleeding out of my neck, so I pressed my finger into my windpipe and tried to stop the blood with my other hand. I stumbled out onto the boardwalk and another captain saw me. He wrapped a sheet around me and took me to the base hospital which was not too far away. I must have passed out because when I woke up, I was in Hong Kong and then Tokyo, Japan. I was there for probably two or three weeks until they shipped me to Travis Air Force Base in California. A couple of days later, in February 1969, they sent me to Fitzsimons General Hospital in Denver where I spent the next year and a half.

At Fitzsimons, we had a bunch of guys who lived in Barracks 402. That's where the bachelor officers lived. There were eight or ten of us there. We had a shared open living area, and everybody had their own room and we were able to go to the officers' club and junior officers' club. So, it wasn't too bad.

Every 30 days I would get to leave after surgery. I had 11 surgeries on my throat. At one time, I even had a colonel ask me to participate in a study. He had been going down to the pound to get stray dogs, shoot them in the throat, and then patch them up. After four days, these dogs all died because they had blood clots. Anyway, he wanted a human volunteer to work on. So, like a dummy, I volunteered. They put me in a main hospital where I was supposed to have an officer of the day (OD) with me all the time.

About in the middle of the night, I started off choking and couldn't get to the OD. Luckily, I had next to me a kid who was all shot up and his mother was with him. She ran down and got the OD who was all the way down in the cafeteria. By the time he got up to where I was, I was on the floor choking. The internal stent had failed. So, after that, I said no more.

After a couple more months of processing, they let me out and I went back home. I received a temporary discharge in February 1969 and a permanent discharge, medical retirement in February 1970. While I was in the hospital, I was promoted and discharged shortly afterwards on medical retirement with the rank of captain.

Before being discharged, I lost 100 pounds after having to take all food through a tube up my nose or my throat. I only weighed 110 pounds. I didn't look like myself. A friend of mine and I went to a basketball game and nobody recognized me. After that, I just reloaded bullets and shot prairie dogs to try and keep busy!

My Letters

August 1968 to May 1969

There's a gap here that was probably a result of me being in transit. I went to Vietnam in August of 1968 and was assigned to the 116th Assault Helicopter Company with an A.P.O. (Army Post Office) address of San Francisco 96353. We were at war, of course, so there was no specific return address and no stamp on the envelopes. We just wrote the word "Free" in the upper right hand corner. Also, not every envelope had a postmark on it and I didn't always write the date so these letters from Vietnam may be out of order, but I'll do the best I can.

An undated letter tells of leaving California with most of my classmates on the night of the 25th which was probably August, I guess. Flew for 19 hours to get to Vietnam. "Spent a nite outside of Saigon, mortars and artillery go off continually, so I guess you could say we are in a War Zone." Our base is 25 miles northeast of Saigon. "Out unit is an Assault Helicopter Co. supporting the 25th Infantry Division and also the 101st Airborne." "We live in a little wood hut (hooch) which everyone fixes up to their taste (some not bad) and all our housekeeping and laundry, shoes, etc. are done by Vietnamese women (Mama Sans) for $9 a month for 5 or 6 in a group. We fly from 6:00 to 6:00 or later depending on the mission. All are combat assaults with the 25th Inf. Div." "I'll be flying slicks (transport helicopters) hauling troops into landing zones on attacks escorted by gunships which I hope I can get later on. We fly 8 to 15 or more ships in formation." I ended up living with Wayne from OCS. "Small world, huh?"

One of my other early letters tells that I'm "Pretty well settled down now and the time seems to be going fairly fast. Getting a lot of flying time in, the days are pretty long." I mention that watches and cameras and stereo set-ups are cheap here and if Dad needs a new camera, I can get one for a lot less than stateside cost. I'm considering a camera to take photos over here and a stereo but probably not until I'm ready to leave. If the folks need money for the furnace, they can go ahead and take some out of my account. Just leave enough in the account to cover checks I might write when I go to the big PX in Saigon. I'm enclosing pictures of what our hooches (Quonset hut barracks) and hooch maids look like. (Again, not sure where those photos ended up.) Got some letters from home including one from a girl who I don't know how she got my address. I'm going to a party today at the officers' club. "They have a bunch of nurses and Red Cross girls around, so I guess I'll have to start looking around when I get the hang of things around here."

No date or postmark on this next letter but I'm talking about flying formations, getting up at 5 or 6 a.m. and get a lot of flying in. I'll send some photos. I need some things from home that I can't get here: my small transistor radio (if you can find it), vitamins, my green cut-offs, special soap, and hair tonic. So this was probably early on.

No date on this next letter but I mention that "Today marked the first month of our tour. Time goes rather fast after all." So, it was probably written in September of 1968. I got my wish, which I thought was almost impossible, to move from flying in formations and picking up troops to starting to fly gunships. I now fly about three times a day compared to ten or eleven times in formation. I'd been flying with my platoon leader in formation in lead position and he asked me if I wanted to go to gunships and I told him definitely yes. "The first day I fired enough rockets at $90 apiece to buy a new 442 or G.T.O." Flying gunships is "one of the most desired jobs and there are only about 9 or 10 out of 50 or so pilots, so it was considered almost unheard of for me to get in with only 3 weeks of experience. Also, (the gunships I now fly) are right in front where all the action is, but they have the protection for it. They had two shot down about a week ago but no one has been hurt in one for a long time. I think they are probably safer most of the time since there have been several guys shot in the slicks (transport helicopters) by snipers and machine gun in the hot LZ's (landing zones) since we've been here. Should be a good year for hunting."

In a letter postmarked October 2, 1968, I'm asking if my folks have my address wrong since I've only received Dad's one letter. I want to know if they got my last check because I'm planning to send another one for $500 more. I bought a $300 camera and all the accessories, tripod, telephoto lens, for $100. I served as pay officer the other day "and handled some $60,000. Sure felt good to see all that money. Right now I'm investigating an aircraft accident for Battalion so it's not all flying. The other day I flew the smoke ship which sets up a protective screen for the fleet. We ended up with a body count of 15 for the guns, 5 for Smokey, so the hunting is pretty good sometimes. Tonight we are expecting a mortar attack and are on standby to go out and hit their positions. Tell you how it works out. When you send me my things, send me a pair of 32-32 Levis. (A month later was when I got hit in a mortar attack.)

I dated my next letter October 4 but for some reason it wasn't postmarked until November 7, two days after I was wounded on November 5. That's the Army for you. I just got through paying everybody and now I'm concentrating on finishing the pool. "I had today off and set the forms for the walks. Wednesday I'm going to pour the concrete and maybe get a crew to put in the ladder. Will send you some before & after pictures of my project. Who would have guessed I would be sent to Vietnam to construct swimming pools. Actually, it is interesting and keeps your mind off flying for a while. Same old story almost every day now. We did get shot up a little bit the other day. My ship only took a couple of hits and the second ship caught fire, but the third ship got a full clip and crashed in some 100-foot trees. Nobody was hurt and after 3 hours on the ground dodging Charlie they all got out O.K. We destroyed the (downed) ship in place with rockets. (It was) one of our best ones." Still planning trips to Australia and maybe Hong Kong and Thailand or Malaysia. Still thinking about extending over here to get the 30-day leave anywhere in the world. Might do that if I don't get Germany or South America for my next tour. "As much as income tax will hit me, I should stay over here. I imagine after a year it could kind of grown on you though." Asking about car sales and land sales back home and hoping my brother is having the time of his life (in college.)

I date this next letter October 6. I'll quote the first part here. "Dear Folks, Finally received your letter, but must admit that it made up for several. I'm sending a write-up from flying Smokey the other day. One of the fellows upset a ship day before yesterday and totaled it out. He had 3 days before he left to go home. No one was hurt but you never know when something like that can happen. I don't think you need to worry as much as I think you probably are. So far I don't mind things too bad over here. We're fixing up our hooch pretty fair, putting a swimming pool in the company and have parties every so often. I'm saving money and staying out of trouble. As far as the flying, we definitely see our share of the action. The Stingers (our gun team) are quite well known over here and go looking for Charlie instead of waiting to be shot at like a lot of the other companies. But so far things don't bother me like some of the fellows. Some are almost nervous wrecks, like caged animals. When we get in some action, it's almost like hunting or playing ball, you have full concentration and determined to succeed. It's not until later that you look back and realize what could have happened. There's no sense in worrying about it. If I did, something probably would happen because I'd be so conscious and nervous that I would make it happen. I consider myself lucky but am anxious to get more experience to prevent some of these accidents from happening to me, or if one does, to cope with it successfully." I go on to talk about how rumors fly around DeBeque pairing me up with one girl or another. "If it wasn't for my family still living there, I would set DeBeque on its ears as far as rumors go." I mention that I'll probably not be able to get to Hawaii, but a buddy and I are thinking of visiting Australia and maybe Hong Kong around the first of the year. I've thought about extending "six months over here for the money, free 30-day leave anywhere in the world and no second tour." I'm growing a moustache (trademark of gun pilots) but it's rough yet, only several days old." Have a lot of letters to answer from Grannie and others.

(Telegrams figure in here. The reader will find them somewhere in this book.)

November 11, 1968, I'm writing on Red Cross stationary from a hospital in Japan: "Dear Folks, This is just a short note to put your minds to ease. I imagine the Army didn't tell you much when they wired you. I'm not going into details in a letter. Our hooch took a direct hit on a mortar attack. I was asleep in bed and took a few shrapnel. I'm in Japan now getting all the medical attention but probably will be sent home to recover at Fitzsimmons in Denver. I'm not sure how long I'll be here, maybe a couple of weeks. Call you when I get to Denver. If any changes, I'll let you know. Don't get all excited like I know you will. I'm feeling pretty good and still in one piece. Love, Mike."

There's a card here in an envelope postmarked November 25, 1968. It was written by the wife of Lieutenant Colonel Albert Rhea Woodruff. He was in Vietnam, assigned to the 269th Combat Aviation Battalion Headquarters at Cu Chi and my unit, the 116th Assault Helicopter Company was under that headquarters. She was in Japan at Camp Drake where I was evacuated to and Lt. Col. Woodruff asked her to check up on me. She did and wrote a note to my parents which must have reassured them. "Just a note to tell you that I've seen and talked to Mike, and he's doing well. My husband called me from RVM (Republic of Vietnam) and asked me to check on Mike and Capt. Gardner—they were wounded at the same time. I located Capt. Gardner at (Camp) Zama and Mike at Camp Drake and was able to drive Capt. Gardner over to see your son. I've checked with the nurses and doctors on the Ward and all assured me he is doing even better than anticipated! His morale is excellent and it's easy for me to understand why my husband admires each of the "Hornets" as he does. They are all marvelous young men, and I feel quite privileged to have met some of them. If there's anything at all that I, or my husband, can do for you all—or for Mike—at any time, please let us know. You have a wonderful and very brave son!"

My letter from Camp Drake, Japan, is postmarked November 23, 1968. Here it is: "Dear Folks, Well, I finally feel like a human again. After the operation yesterday I'm eating food again, after 2½ weeks of tube feeding. I'm at 150 lbs if you can imagine after my 180 and 190 in A.I.T. (Advanced Individual Training) 1967. My voice is still pretty weak but the Doc said the vocal cords looked O.K. Also I'm walking around with a cane but I can do without it. Surprising how much the human body can recover in a couple of weeks. I should be in Denver around the 1st of Dec. sometime. From some of the cards I get I believe some people might have a misconception of my injuries. They say how lucky I am. Maybe they mean to be alive, which is right. I know the Army doesn't tell you much, probably a good thing to keep you from worrying. But from the way I understand it, I was on the critical list until after I got here. Now that I'm O.K. I might as well give you an idea of what happened. Flying gunships is considered to be one of the more dangerous jobs over there right now except the ground troops who I have the utmost respect for, and sympathy. Well, anyway outside of a few hits in a lot of action which I ate up nothing happened to me. The night of the 5th I stayed up until about 12:00 thinking about some work on the swimming pool since I had the next day off from flying. Wayne was in Saigon, a good thing I guess from the description of his bed he would never have made it. Anyway about 1:30 A.M. our hooch took a direct hit from a mortar. The Club took a hit but luckily it was closed. I think there were a total of 13 rounds on post. The 1st time since around July. I was knocked out of bed and I crouched under the bed not knowing I was hit. It just kind of numbed me. I started gasping for breath (I thought from the smoke) and could feel some blood. I got up and looked for our other roommate on the porch but he was gone. By now I was having real trouble breathing and bleeding pretty bad. I finally realized I couldn't breathe through my mouth so I open a hole in my windpipe and got air that way. I say lights on the captain's hooch about 50 feet away. By this time the post was alerted, sirens going and most people in the bunkers. Luckily he was still on the floor. He got a medic, and I wrapped a sheet around my neck to stop the bleeding. I still have a metal thing in my neck to breathe through. Quite a few guys do with broken jaws etc. After I got out here I found out I had 18 holes the size of a quarter or bigger and 3 about the size of a dollar bill. No real damage though. The small fracture of my left ankle and some shrapnel still in me here and there. There is a piece by my left lung that they don't seem to be concerned about. Maybe after I get to the states they'll look into it more. Now you see why I'm coming home. Most of the wounded go straight back so it's not a way of getting out of your tour. I was going to wait until I got home to go into details but some of the letters set me to wondering. (He has a sprained ankle but he got out of Vietnam.) Personally I would just as soon stayed like I was. Funny, now I'm not too set on going back if I have to. Kind of makes you wonder if somebody is trying to tell you something. If I had got shot down it would have been different. I would have been anxious to get back because you expect it. Kind of like getting bucked off and getting back on. But when you're sleeping? Oh well, a lot of time to think about it. Well, so much for the War Stories. Right now I'm feeling spry. Should be home by the time you get this. Let you know as soon as I get to Denver. Should be able to go home for the holidays. At least I hope so. Could take a big drink right now. Going to set some high calories foods and with the weight I should be back to my weight in a month or so." I blanked out some things I wrote about warning Perry about some girl, so best just leave that out. I closed this letter with a note about a guy from DeBeque who I kept running into in basic training and again at flight schools in Texas and Alabama. "He got shot in the neck and had the same thing in his throat. He left for Denver today and will probably still be there when I get there. Small world isn't it?"

In a letter postmarked January 27, 1969, I show a return address of an apartment on Pennsylvania Street in Denver. There may be some letters missing because I got transferred to Fitzsimmons sometime in there. My January letter is mostly about money. My back pay and pay records haven't arrived from Vietnam. Anyway "I saw a specialist today and he showed me what they could put in and maybe help (my throat, I guess.)

Tomorrow they are all going to get together and discuss all possibilities, so maybe I'll know more. They have to put me to work now that I'm out of the hospital. In the afternoons I go to the gym for reconditioning." I'm working out with weights and basketball. I'm scheduled for five days but I can only manage three or four. Fridays I see the doctor. "I can't sit for long periods and can't talk all day so I don't know what I'll do. The specialist has some dogs for experiments and said I could go out 3 half days a week and work with them." I'm thinking to try flying fixed wing aircraft that might be a nice set-up if I can arrange it and I could do that instead of other work. For now I'm working at a clinic to keep from getting bored. May come home to DeBeque for the weekend. Will call if I decide to.

By late May 1969 looks like I'm back in Fitzsimmons because the return address is a hospital ward there. They are going to look down my throat while I'm asleep and supposedly get a full picture of what it is like. I should know something this week about taking a leave or moving out of the hospital. I've got a car which I've had fixed up and I need some paperwork sent to the dealer. I'm hoping Perry can sell my motorcycle because I don't want to keep paying insurance on it. The problem with taking a 30-day leave is money. "Next month should be a better month after

I make captain. 20 days left." I'm thinking about a trip to Georgia if I can get someone to drive with me. Anyway, "I'll probably be home around the 1st if not sooner, if everything works out."

Oddly enough, my re-discovered letters end in an envelope postmarked May 29, 1969, with three separate letters crammed inside. No dates so not sure of the order but the topics are cars (of course), another operation, and finances (what else?) Let's look at the operation first. "Well, I guess the operation was a success. At least if they cut on anything I didn't know it. They talked like they were satisfied with the way it looked. I was still drugged but I thought I heard them say something about going back to duty maybe at Ft Carson because it is close to the hospital. I guess after my leave I'll have to push for a discharge. It seems like the only way to get one." The Olds dealer is still bugging me about the paperwork. I ask my folks to send it as quickly as possible. Need to get some money in the bank. "Doc said I looked better this morning but that they'll keep me here for a while. Looks like they are going to operate, but I don't know when. I'm going to try and get out before they do. He told me that I might as well get rid of my apartment since I wouldn't be there much." That will put the other guys (my roommates) in a bind. My mother may come over now or wait until the surgery. I'll try to pin them down here on some definite dates and call when I find out more. I must have the surgery. "I got out Monday as Mom has probably already told you. I was quite disappointed at first at the results. But now it seems like it may be coming back a little. I'll try to come over in the next couple of weekends when it gets nice. My tires aren't as good as I thought they were." More about the car paperwork and cost of insurance and asking about my land and whether I might be able to sell it. I close what appears to be the final letter in this bunch with "Well, I'll try to call after I get to talking a little stronger. Love, Mike."

Laying smoke to protect a landi
zone

Flying in formation

Army hole, Navy concrete: buil
the pool I never got to see. The
mortar attack which wounded m
hit my hooch which looked like
one in the background.

Recuperating at Fitzsimmons Army Hospital in Denver. I'm the one with a gun pilot moustache.

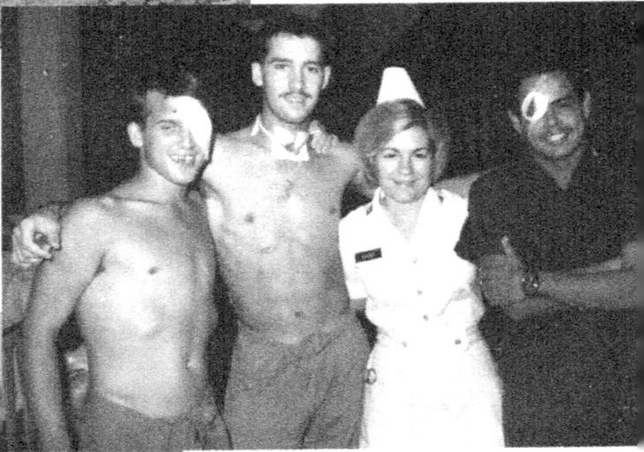

Speech therapy at Fitzsimmons

CLASS OF SERVICE
This is a fast message unless its deferred character is indicated by the proper symbol.

WESTERN UNION
TELEGRAM

SYMBOLS
DL = Day Letter
NL = Night Letter
LT = International Letter Telegram

The filing time shown in the date line on domestic telegrams is LOCAL TIME at point of origin. Time of receipt is LOCAL TIME at point of destination

539P CST NOV 6 68 KA461
SYB377 SY WA412 XV GOVT PDB 4 EXTRA FAX WASHINGTON DC 6 532P
EST
VERNON LEO HOTZ, DONT PHONE, DONT DLR BTWEN 10PM & 6AM, CHECK
DLY CHGS ABOVE 75 CENTS
DE BEQUE COLO

THE SECRETARY OF THE ARMY HAS ASKED ME TO INFORM YOU THAT YOUR SON, LIEUTENANT MICHAEL V. HOTZ, WAS WOUNDED IN VIETNAM ON 6 NOV 68 AS THE RESULT OF HOSTILE ACTION. HE RECEIVED A CRUSHED LARYNX AND METAL FRAGMENT QOUNDS TO HIS FEET, LEGS, AND BUTTOCKS. HE WAS IN HIS BILLETS IN BASE CAMP WHEN THE CAMP CAME UNDER A HOSTILE MORTAR ATTACK. PLEASE BE ASSURED THAT THE BEST MEDICAL FACILITIES AND DOCTORS HAVE BEEN MADE AVAILABLE AND THAT EVERY MEASURE IS BEING TAKEN TO AID HIM. HE IS HOSPITALIZED IN VIETNAM. ADDRESS MAIL TO HIM AT HOSPITAL MAIL SECTION, APO SF 96381. YOU WILL BE KEPT INFORMED OF ANY SIGNIFICANT CHANGES IN HIS CONDITION. PLEASE TELEGRAPH COLLECT, CHIEF, CASUALTY DIVISION, DEPARTMENT OF ARMY, FAX, WASHINGTON, D.C., THE NAME AND ADDRESS OF HIS MOTHER

KENNETH G WICKHAM MAJOR GENERAL USA C-682 THE ADJUTANT GENERAL

KA054 K DVE065 1968 NOV 18 PM 1 54
(L WA047) XV PDB 4 EXTRA=FAX WASHINGTON DC 18 236P EST=
MR AND MRS VERNON L HOTZ, DONT PHONE CHECK DLY CHG ABOVE 75
DEBEQUE COLO

=ADDITIONAL INFORMATION RECEIVED STATES THAT THE CONDITION, PROGNOSIS AND MORALE OF YOUR SON LIEUTENANT MICHAEL V. HOTZ ARE GOOD. ITS CONTEMPLATED HE WILL BE EVACUATED IN APPROXIMATELY THREE WEEKS. YOU WILL BE PROMPTLY ADVISED AS ADDITIONAL INFORMATION IS RECEIVED=

KENNETH G WICKHAM MAJOR GENERAL USA C-2277
THE ADJUTANT GENL=

855P CST NOV 27 68 KA741 NSA645
NS WA115 XV GOVT PDB 4 EXTRA WASHINGTON DC 27 NFT
MR AND MRS VERNON L HOTZ, DONT PHONE, CHECK DLY CHGS ABOVE
75 CTS
DEBEQUE COLO

ADDITIONAL INFORMATION RECEIVED STATES THAT THE PRESENT CONDITION, PROGNOSIS, AND MORALE OF YOUR SON, LIEUTENANT MICHAEL V. HOTZ ARE GOOD. HIS PRESENT DIAGNOSIS IS MULTIPLE FRAGMENT WOUNDS TO THE NECK AND EXTREMITIES, LARYNX INJURY, RETAINED FRAGMENT IN THE LEFT LEG. HE IS DOING WELL. ALL TUBES ARE OUT AND HE IS ABLE TO TALK IN A VERY POOR VOICE, WHICH SHOULD IMPROVE. ALL HIS WOUNDS ARE CLOSED. EVACUATION TO THE UNITED STATES IS CONTEMPLATED WITHIN 2 WEEKS. I SHARE YOUR HOPE THAT HE WILL HAVE A RAPID RECOVERY

KENNETH G WICKHAM MAJOR GENERAL USA C-3591 THE ADJUTANT GENERAL
(900).

SF1201(R2-65)

Chapter 4

Colorado State University

1969 to 1970

After getting tired of shooting prairie dogs for a month or so, in the autumn of 1969, I ran into my friend's older brother. He'd just gotten out of the army. He served in Vietnam with the Big Red One Division, and he had carrot-colored hair. So, I called him red on the head (I meant it as a joke, like dick on a dog.) But he didn't like that name very well. Anyway, we decided to go to Colorado State University. I don't know why. So, we enrolled.

We were having trouble finding a place to live, so we ended up staying in the basement of a motel. It was a pretty nice place, out of town a bit. Being a freshman, I was about 21 or 22 and everybody else was 18. So, it just wasn't my league. I couldn't get used to it.

So, Lyle from the old days was back, though not in school with me. And he and I would meet downtown and go to bars and drink red beers.

We took some pretty easy classes. I took a real estate class, business, golfing, bowling, and shop class—anything just to get some credits. I did okay on those easy classes, but I also took flight lessons in a fixed-wing aircraft. But, after ten hours, I figured out it wouldn't work. Between my voice which wasn't clear, and dealing with a different flight picture, that is flying using instruments alone, I figured I'd better stop.

I didn't have much interest in what I was doing in school, so I would go to the dog races instead. But the racetrack was far away, and I had traded my 442 on an old Ford pickup which I didn't much trust. So, I rode with the owner of the trailer park, where I had moved. I made some money there when I figured out a numbers system which worked for me, especially if it had rained and the track was wet. The trailer park owner started getting together with a couple friends and they thought I knew more than I did. I didn't know that much, although I just missed winning big a couple of times when the odds were crazy, like five to 6,000 or some such impossible number.

Anyway, after only a year, I left CSU and when Lyle and I got back to DeBeque, we started partying again. For work, we started hauling railroad ties and fence posts as just a way to make some money. To do these odd jobs, I had to borrow Dad's stock truck and then get it back in time or else he'd get mad. Sometimes I would just be out all night around the town and go to a Grand Junction party, so my time wasn't very constructive. But Lyle and I just had to get some stuff out of our system, I guess.

One time another friend Bobby and I went up on the mountain and stayed 30 days. We met some hunters there and we let them walk off of Grand Valley Point while we sat waiting for them in the bottom. We started drinking whiskey and went over across the highway to see a neighbor and, yes, he had young daughters. Well, anyway he used to have some daughters when we were all young, but those females had moved off to go ahead with life. Anyway, Bobby and I ended up shooting three big bucks and then we went to a bar in Grand Valley and got drunker and somebody must have heard Bobby talking. Because, when we went to get the deer later that night, a game warden was waiting but he got in a wreck in his truck, so we weren't fined then.

Later, about midnight, the game wardens and state patrolmen came out to where we were staying trying to get confessions out of us. We didn't confess but we decided we'd better quit the area. I went to Gunnison and Bobby went to Durango and I think he got married right away out there. I got married later in Grand Junction, in 1972, but that didn't last a year.

In the time I was in Fort Collins where CSU is located, in addition to taking classes, I was working all the time in real estate. I had bought a small house right on campus, also a trailer out of town and a new tri-level building. Lyle and I stayed around for a bit to sell the house.

At that time, Lyle had a girlfriend from back home who was teaching about 30 miles away. I remember she had to take some kids on a field trip for a couple days. I don't recall exactly how all this happened, but one night, while the girlfriend was still out on the field trip, we had a party at her house. The party lasted nearly all night. Then Lyle got a phone call from his girlfriend saying one of the students had gotten sick, so they were coming home early. We had to kick everyone out and clean the place up only to discover the couch had a big hole burned in it. I left Lyle to figure out what to do and I never heard how things were resolved. I just assumed Lyle somehow made amends.

I got started in real estate and cattle—two enterprises which kept me going for years.

Chapter 5

Crested Butte and Gunnison

1970 to 1982

After CSU and time in DeBeque, I went over to Gunnison and also to Crested Butte. My Dad and brother were building houses up there towards Crested Butte. So, I stayed there for a couple days a week until I saw that more was going on in Gunnison. Anyway, one night my brother and his workers wanted to go to a bar down at Crested Butte. I didn't feel like going but they finally talked me into it.

Dad had a neighbor whose daughter was working with my mother cleaning houses. And it was decided that me and this girl would go down to Crested Butte. I didn't want to stay very late, and these guys were getting drunk, so I went with that gal and ended up having a thing with her. But Dad showed up and got mad about us being together. So, I drove back to Gunnison that night.

At the time, I had that old Ford pickup I had traded for. I hadn't bought anything new yet. I drove home and then I drove back again and ended up going back to another bar later in Crested Butte. This was when the Frazier and Liston fights were going on and the bar was showing the fights. I ran into an old guy that was drinking shots at the bar. Turned out he owned the bar, and he was thinking about selling it. So, I told him I was getting in the real estate business, and he said he would list the bar with me. So, I got the sale listed for $110,000.

Then, on a Saturday, some guys came through town from Texas and bought the bar, so I closed on it. I was working for United Farm Agency and so I listed it sold. When I got my check, it was $1,500 out of $10,000. I said, well, that's not going to work since I had to pay some to United Farm. So, I decided I'm going to go work for myself.

I opened up an office downtown with Bud Roberts in Gunnison with a real estate company named Safe Buy and I got a partner. We went ahead and decorated the office and started taking listings. One of the reasons I did all this was Dad lashing out at me because he was upset that I was messing with his neighbor' s daughter. He told me that the army hadn't done me any good. That must have struck a bell with me, so I decided to start doing something.

Anyway, we started getting a lot of listings. Started making some money on 100 lots by Crested Butte that we bought for a thousand a piece and sold at $10,000 each—$230,000.

During this time, I met a guy down at the bar at Cattlemen's Inn and he was looking for a partner on a ranch. I heard you do a lot of activity, he said. I don't have that kind of money for a ranch, I said. He said we could get financing. He also told me I'd do the cattle and run the ranch and oversee the work of the Mexicans. So, it sounded to me like a good deal.

This guy and I finally got it done through Connecticut General Insurance. That changed my work with real estate. I turned it over to Bud Roberts and I came in occasionally but was too tied up at the ranch to do much. I had 1,300 acres of home ground and 50,000 acres of range under permits. Ran about 800 cows. We started out running yearlings. Brought some in from Arizona and neighboring sale barns. I ended up with about 1,176 head as I remember. So, we hired a full time range rider, or summer range rider, who each day kept our cows away from poison plants and water we had up in one area. So, we had a very low poison rate, like maybe ten out of that many head. That rate was unheard of.

Then, come fall we put so many head out on the market that we had a great sale. We started getting a lot of buyers coming in. I remember one guy that had my Dad farming the ranch back home came up and told me about what a wild dude I was. I said that may be but if you want to buy our steers, make me an offer. He thought I was wild because I'd got his new hat wet crossing ditches, and they joked about that.

Anyway, I ended up selling to a guy from Greeley and he took out four semi truckloads in the first draft after we worked them up and done alley ways by noon. I don't bet things would have gone better but one of my men wasn't working fast enough so I switched with him. He went to the scales, and I went to the alleys, and I just told our buyer that loading was causing a lot of shrink on his cattle. (Meaning the cattle were losing weight during loading which sometimes happens when Nature calls.)

So, I told him, we're going to load better. You stand back. The stock is all going on the trucks, but the Mexicans will pull out the Brahmas. I wanted to sell those for a different price. There were only like a dozen Brahmas, so he didn't like it, but he agreed.

Well, after the first truck went, he didn't give me a draft for payment. He said it was late. Later, he said he would pay with a second draft which again he didn't pay. Then, a couple of weeks later, he pulled the same thing. So, I wouldn't let him leave town until he had a draft for me to pay for all. It took until about dark to do all this and I had to keep the banker at the bank until the draft came through.

He had wanted to move the trucks earlier. But I had a couple renters who worked at the Gunnison Police Department. I decided to see if they'd put on their uniforms and go sit by the edge of town and not let the trucks leave until I got paid. So, the trucks sat, and the cattle did what cattle do, and he said I'm getting a lot of shrink. I said, well what goes around comes around. He said, next time I'll bring cash and I said, there won't be any next time.

By that time, I was getting pretty hard set in my ways which came from running the ranches and running with the old cattle buyers. Anyway, it ended up a pretty good season.

Chapter 6

Ranching and Deena

1982 to 2025

Since I was working the ranch and also at the real estate office part-time, I didn't have a lot of time for my personal life, so my wife at the time ended up going to Junction. Her folks live down there. Anyway, she started running around so when she didn't come back, I looked at that relationship as probably over. Sure enough, I got divorced.

My brother, Perry, had the bar next to my office. At the bar, he met a little cocktail waitress named Terry and her roommate was Deena. So, I started hanging around Deena and ended up later and down the road married to her. Perry's still married to Terry. It's been 53 years.

So, Deena moved with me out on the big ranch in a big brick house on a hill property overlooking the Hay Meadows. And she ran a travel business in my other building. Her mom was also a travel agent. Trouble was they traveled all the time and when it came time for branding or haying, hunting or whatever she was gone. So, I had to get one of the other wives to take care of feeding chores and getting supplies. Which was hard seeing as we were 30 miles out. It wasn't like you could just order a pizza.

I took a year to find some other ranches and I took the office help and ranch help on a trip over to Hawaii. I wanted everybody to have a vacation for almost a week.

While on that trip, as I was running through the airport, I signed papers at 18 percent to buy my helicopter. When we got back from Hawaii, my wife decided she wanted to build a new house on the river on 27 acres I had bought. Beautiful piece of property. But I said, no we're not going to build a house yet. I need to buy a couple of other ranches. Well, that started an argument.

"So," I said, "it has been raining for two weeks and I'm going to go hunting up in Alaska."

She said, "You go to Alaska, and you'll be divorced when you get back."

LOL, I thought, *probably not.*

"I'll be gone around 30 days," I told her. "So, sure, whatever."

Anyway, I left not knowing whether I'd still be married when I got back. All this took place in 1976. Alaska was not exactly a picnic. Neither was getting back home.

So, another divorce happened but I did get a moose, bear, wolf, and caribou. It was a great trip although for part of it we got stranded up there for eight days. One Indian guide and one helper didn't bring the horses in when they were supposed to. Eventually, we got to a lake where the seaplane would come in and they ended up taking us back to British Columbia where we were able to fly back to Salt Lake and then to Denver. We had to get all our meat and horns back plus our guns and equipment so that was a several days adventure getting back home.

So, all that lasted about seven years and what it came down to was I had a chance to sell the ranch. I didn't want to sell it, but I didn't have enough money to buy my partner out and, in the meantime, Homestake Mining came along to take over the water rights to be shipped over to Denver.

We had bought the ranch for a million-three, and they offered three-and-a-half million. Of course, my partner wanted to get out and since I couldn't buy him out, I didn't want to pay the taxes. That would have cost me over $500,000 in tax money. So, I said I was going to do a trade.

Now, before this all happened, my partner was up in Crested Butte doing cocaine at the bars and Federal Land Bank and PCA, production credit was trying to cut us off our financing. Since that could happen, we didn't have the money to operate by ourselves. So, I told my partner I wanted to end our partnership which was another reason to sell.

One time, before all this business of selling the ranch happened, my partner brought some cocaine down and wanted me to try it. I tried it but I said, well, it's like drinking a bunch of coffee. Later, I told him you have got to stop cocaine, and we've got to stop being partners because the finances are going to quit us. And even if we lose our financing, your reputation will follow us anyway.

Sometime, while all this was going on I seem to recall an odd memory. One of my ranch hands was doing my horse shoeing and I had decided to go to the mountains to hunt wild pigs. This must have been in Hawaii, but the memory is hazy. Anyway, I somehow managed to get my feet sunburned so bad that I had to buy a pair of size-14 lace-up boots to even be able to go hunting.

Anyway, the ranch got sold and I started looking for other ranches to buy: Tomichi Dome, Razor Dome, and Juanita Hot Springs. I bought those and used up my share. I had to restock all the animals, except horses which we had kept. I acquired 30-plus buffalo. Ended up with a little more stock and no partner. I felt good. Thought I couldn't go wrong. Had the golden touch.

I bought newer and bigger equipment, including my own semis and cattle trailers. (A new truck cost ten thousand then, now it would cost a hundred.) My properties became the biggest ranch in Gunnison. Meanwhile, the divorce cost me plenty: apartments, hotel, trailer park, houses, and cars. But I didn't care. I had no partners at all.

The last financial statement I've found from that time showed I had $5.3 million. Go figure.

Then the economy in Western Colorado fell. I had to sell the helicopter, which was the first one used to gather wild horses on the Bookcliffs. Then I sold the semis and dozers, some older cattle, and two ranches. Things ended up in several lawsuits on foreclosures.

I started going to Grand Junction with my big horse trailer to trade some cattle. I also went to the big city so I could catch some bars. Brought some city girls who wanted to play cowgirls back to the ranch, and I had kept my big boat in Lake Powell, a 28-foot cabin cruiser.

All in all, I've had four marriages and four divorces. I haven't mentioned all the details but, thinking back, I've decided maybe divorces weren't the way to go. Instead, I should have stuck with buying land, horses, and cars.

Cows and horses and land and kids and cars amd hunting and fishing—
that pretty much sums up my recent years.

CHAPTER 7

What's Next?

There's more yet to tell but I wanted to get this first book out as soon as possible. You never know, right? I'd like to say more about the ranches and my land and my cars. And more about the family too and where I've ended up. We'll see. At least I made it this far and mostly by grabbing the horn as it came.

On my hunting trip to Alaska, I shot a certain mountain goat which I had mounted on the wall of my Fruita home. One day, while pushing my way from room to room in my wheelchair, I managed to knock the mounted head down and the old goat almost speared me to death.

It's a good thing I survived, otherwise somebody would have had a good joke to tell about how old Mike killed that goat from more than 625 yards away and the goat got him back from three feet.

Speaking of Fruita, I might as well mention how I happened to end up in a wheelchair. I was drinking on Christmas Day 2020 when I fell backwards down my steps. I knocked myself out cold and broke five or six vertebrae in my neck. That little adventure earned me six months in a hospital bed.

Nowadays my health is good, but I still use a lot of medical equipment to get by.

Made in the USA
Monee, IL
04 November 2026

67378974R00056